FOSSIL FUELS

IAN GRAHAM

WAYLAND

ENERGY FOREVER?

Fossil Fuels

OTHER TITLES IN THE SERIES
Solar Power · Water Power · Wind Power
Nuclear Power · Geothermal and Bio-energy

Produced for Wayland Publishers Ltd by
Lionheart Books, 10 Chelmsford Square, London NW10 3AR.

Project editor: Lionel Bender
Designer: Ben White
Text editor: Michael March
Picture research: Madeleine Samuel
Electronic make-up: Mike Pilley, Radius/Pelican Graphics
Illustrated by Rudi Vizi

First published in 1998 by Wayland Publishers Ltd
61 Western Road, Hove, East Sussex BN3 1JD

© Copyright 1998 Wayland Publishers Limited

Find Wayland on the Internet at http://www.wayland.co.uk

British Library Cataloguing in Publication Data
Graham, Ian, 1953-
Fossil fuels
1. Fossil fuels - Juvenile literature
I. Title
333.8'2

ISBN 0 7502 2229 8

Printed and bound by G Canale & C.S.p.A.,

All Wayland books encourage children to read and help them improve their literacy.

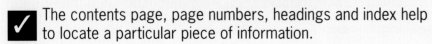 The contents page, page numbers, headings and index help to locate a particular piece of information.

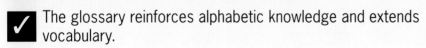 The glossary reinforces alphabetic knowledge and extends vocabulary.

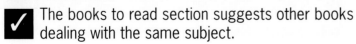 The books to read section suggests other books dealing with the same subject.

Picture Acknowledgements
Cover: Tony Stone/Getty Images/Nadia Mackenzie. US Department of Energy: pages 4 top, 16, 20, 43. Ecoscene: pages 15 (Gryniewicz), 24 (Erik Schaffer), 30, 30-31 (Sally Morgan), 32 (Gryniewicz), 36-37 (Alan Towse), 38 (Sally Morgan). Wayland Photo Library: pages 1 (BP), 18 (BP). Canada House: pages 6, 20-21. Mary Evans Photo Library: pages 8, 11, 12. Samfoto: pages 9 (Ragnar Frislid), 16 (Svein Erik Dahl), 20 right (Svein Erik Dahl), 23 (Hans Hvide Bang). Eye Ubiquitous: pages 13 (Steve Lindridge), 16-17 (David Cumming), 25 (Tim Hawkins), 28 (Davy Bold), 33 (J.B. Pickering), 37 (Paul Seheult). e.t. archive: page 10. Forlaget Flachs: pages 26-27 (Enequist Kommunikation), 35 (Olë Steen Hansen), 41 bottom and 43 bottom (Olë Steen Hansen), 41 top (D.O.N.G.). Shell Photo Service: page 29. Nissan Motors: page 34. National Coal Board: page 42. Science Photo Library: pages 4 bottom (Crown Copyright/ Health and Safety Laboratory), 5 (Simon Fraser), 43 top (Astrid & Hanns-Frieder Michler), 44 (Martin Bond), 45 (Martin Bond).

CONTENTS

WHAT ARE FOSSIL FUELS?

Introduction

Most of the energy we use at home and at work comes from fossil fuels – coal, oil and natural gas. The petrol and diesel oil that road vehicles run on and the kerosene that fuels aircraft are all fossil fuel products. The electricity we use every day is made in power stations, most of which burn fossil fuels. Coal, oil and gas all contain hydrogen and carbon, so are also known as hydrocarbons. They are called fossil fuels because they formed in the Earth's crust millions of years ago.

Will fossil fuels last for ever?

Fossil fuels are a non-renewable source of energy and won't last forever. The energy they provide for us will have to be replaced by energy from alternative sources such as solar, wind or wave power. While fossil fuels last, they have one major problem. Burning them creates air pollution. The poor air quality in many cities is caused by pollution from traffic.

An oil-drilling derrick towers over the well-head at Elk Hills, Alaska. The Elk Hills field produces 200,000 barrels of oil a day. (Oil is measured in barrels. One barrel is equivalent to 159 litres.)

Left: In an underground coal mine, a worker operates a coal-cutting machine while a technician measures the noise level. The men wear protective helmets and ear defenders. Hydraulic roof supports prevent rock from falling out of the ceiling while coal is being cut.

White plumes of water vapour drift skywards from the cooling towers of a coal-fired power station. The Ferrybridge power station in Yorkshire, UK, generates over 2,000 megawatts of electricity from coal.

Where do fossil fuels come from?

Coal, oil and gas come from plants and animals that lived on the Earth long ago. About 300 million years ago, there was a great increase in the numbers of plants on Earth, especially trees and huge ferns. Many plants lived on swampy ground and when they died they fell into the swamp. But they did not rot, as there was too little oxygen in the swamp. They piled up. Eventually, a thick layer of dead plants became buried. As more soil and plants piled up on top of them, the dead plants became compressed underground. First, they formed peat. Then, as they were compressed even more, they formed a brown coal called lignite. Eventually, with still more pressure from above, they formed hard, black coal.

Billions (thousands of millions) upon billions of microscopic plants and animals lived in the seas. When they died in their billions, their bodies sank to the seabed. There they became buried and compressed in the same way as the plants on land. But beneath the sea, they formed a thick black liquid – oil. The oil was trapped underground by rocks that formed from the sediments above it. Gas given off by the decaying organisms could become trapped underground too. This is how oilfields and gas fields formed.

FACTFILE

Lignite, dark brown in colour and formed from compressed peat, is the youngest coal. It may be less than one million years old. When lignite is compressed deeper underground, it forms a soft coal called bituminous coal. When this is compressed even more, it forms the hardest and oldest coal, anthracite, which may be up to 400 million years old.

Left: An enormous power shovel digs out tar sands at Fort McMurray in Alberta, Canada. Tar sands contain bitumen, which is mixed with steam and water to make oil and gas. Canada has the world's biggest tar sand deposits. Fort McMurray produces up to 130,000 barrels of oil (20.7 million litres) from tar sand every day.

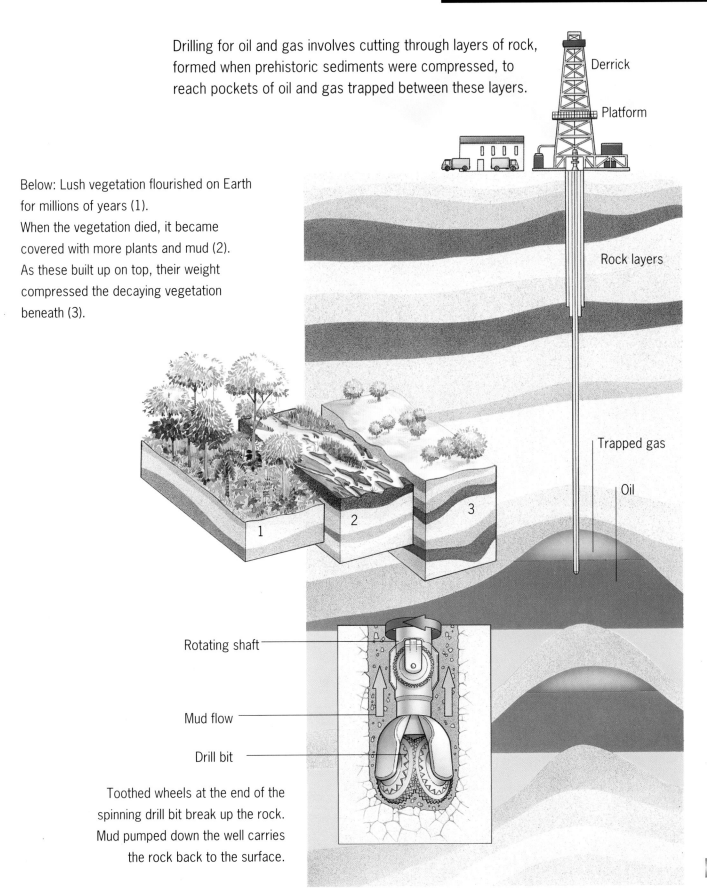

Drilling for oil and gas involves cutting through layers of rock, formed when prehistoric sediments were compressed, to reach pockets of oil and gas trapped between these layers.

Derrick

Platform

Below: Lush vegetation flourished on Earth for millions of years (1).
When the vegetation died, it became covered with more plants and mud (2).
As these built up on top, their weight compressed the decaying vegetation beneath (3).

Rock layers

Trapped gas

Oil

1

2

3

Rotating shaft

Mud flow

Drill bit

Toothed wheels at the end of the spinning drill bit break up the rock. Mud pumped down the well carries the rock back to the surface.

FOSSIL FUELS IN HISTORY

How long has coal been used?

The Chinese may have used coal to make copper about 3,000 years ago. The Greeks used coal seven centuries later, and so did the Romans who came after them. We know that ancient Britons were already digging coal long before the Romans invaded in 55 BC, because ancient flint tools have been found embedded in coal seams. About 1,000 years ago, Hopi Indians in the United States used coal to bake pottery.

In the age before machines, mining was even harder work than it is today, as this illustration of a mine in 1885 shows. Coal was dug out by miners using hand picks and shovels. It was also very dangerous. Collapsed roofs, fires and explosions were common.

The steam locomotive was one of the world's most important inventions. It changed history. It made long-distance travel possible. Steam locomotives transported raw materials to the new factories of the Industrial Revolution in the nineteenth century. They carried people and goods across North America where new towns sprung up all along the railway tracks.

In places where there are underwater coal seams near land, the sea can break up the coal and wash it ashore. Sea coal, as this is called, was collected as long ago as the thirteenth century. Coal was collected from the surface of the land too. When all the surface coal had gone, people dug down to collect more. But mining coal was extremely dangerous. Mines often flooded with water or filled with deadly gases.

The use of horse- and steam-power

Early coalmining was all done by hand. Then, in the seventeenth century, small horses were taken down into the mines to pull carts full of coal. These pit ponies, as they were called, lived underground. Many of them never saw daylight until they were too old or too ill to work any more. In the nineteenth century, steam engines were used in mines to pump out water and hoist the coal out.

The history of oil

Oil has been used in one way or another for about 8,000 years. Oil seeping out of the ground was changed by sunlight and the weather into a thick tarry substance that has been used since earliest times for building, waterproofing and medicine. The Egyptians used it to preserve their dead as mummies. The Chinese used it for heating, lighting, cooking and making bricks. Native Americans used it for heat, light and medicine. In Europe, oil was used to make ointments for treating back pain, bruises and swellings.

Most of the oil used in ancient times was collected from natural seepage at the surface, but drilling for oil can be traced back to the second century BC in China. Bamboo pipes and bronze tubes were used to bring the oil to the surface. Oil was also found by accident when wells were sunk to look for brine (salty water). People found uses for the oil and then deliberately began drilling for it. The modern oil industry began in the United States in the nineteenth century.

The bodies of important ancient Egyptians were preserved after death. Part of the preservation process, called mummification, sometimes involved painting the skin with oil. This gave the mummy a black appearance, as here, where the ribs and the legs are shown.

FACTFILE

Ancient Chinese oil wells were sunk quite differently from modern wells. Today, a rotary (spinning) bit drills itself down into the ground. The ancient Chinese method was to repeatedly raise and drop a heavy chisel-shaped cutting tool. Although the Chinese managed to sink wells as deep as 1,000 m in this way, it could take several years to sink one well.

The first modern well drilled deliberately to reach oil was sunk by a retired railway guard, Edwin L. Drake, in Pennsylvania, USA, in 1859. The drill was powered by a steam engine. The well was 21 m deep and it produced 1,600 litres of oil every day.

11

Above: A lamplighter lights a gas street lamp in 1867. Every street lamp had to be lit by hand at night and turned off again after sunrise.

Gas in history

Natural gas was used as long ago as 900 BC in China. People sunk wells to look for brine (salty water), but they found natural gas there too. They reached a depth of 1,460 metres. The gas was burned to evaporate the water, leaving salt behind. By AD 900, gas was being sent through bamboo pipes for heating, lighting and cooking. In 1664, John Clayton discovered a pool of natural gas near Wigan, England. In 1739, Clayton also succeeded in producing gas from coal. The production of coal gas began in earnest in about 1800. It was piped into people's homes and used for cooking, lighting and heating until natural gas was available in large enough quantities, which took place in the last fifty years.

● Coal deposits
● Oilfields
● Natural gas deposits

Fossil fuel deposits

Natural gas is usually found wherever oil is found, so many of the world's oilfields also produce gas. In addition, large gas fields have been discovered in recent years in the North Sea and in Russia, Canada, Australia and Algeria.

Surplus gas burns from an offshore drilling platform. The flame, or flare, is a safety device that allows any surge in gas pressure to escape harmlessly.

Gas lighting

Gas lighting was introduced in 1792 when the English inventor William Murdock used coal gas to light his home in Redruth, Cornwall. Until then, homes had been lit by oil-lamps and candles. Outdoors, it was pitch black after sunset apart from the few days every month when, if there were no clouds, the moon was bright. Gas street lighting changed that. In 1807, Pall Mall, a main street in London, England, became the first street in the world to have gas lighting.

HARNESSING FOSSIL FUEL POWER

Looking for fossil fuels

Before fossil fuels can be used they have to be found. At first, looking for coal, oil and gas was very hit and miss. The only way to find out if there really were fossil fuels under the ground was to sink a mine or a well. Nowadays, the search for fossil fuels is more scientific. Experts map the ground and what may lie beneath it in a process called surveying.

Scientific surveying

Geologists look for tell-tale types of rocks where fossil fuels may have formed. They study fossils embedded in the rocks. Aerial photographs show the shapes of the surface rock layers. Measuring the magnetic field of the area from an aircraft gives clues about what types of rocks lie under the surface. Measuring gravity shows how dense the underground rocks are. Gravity is stronger above very dense rocks. Setting off explosions in the ground and studying how the sound waves are reflected gives more clues about the underground rocks and layers. Test bores may also be drilled. Samples taken from the test bores and readings from instruments lowered down the bores give important information about what lies underground.

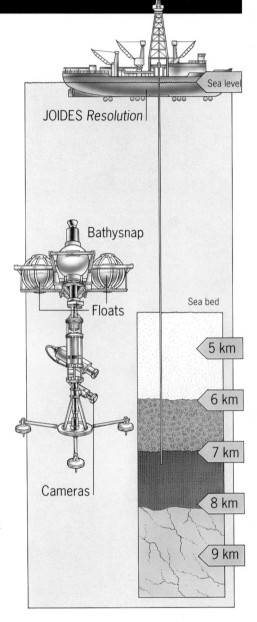

JOIDES *Resolution*

Sea level

Bathysnap

Floats

Sea bed

Cameras

5 km
6 km
7 km
8 km
9 km

Test bores in the sea are often drilled by a specially adapted ship. The JOIDES (Joint Oceanographic Institutions for Deep Earth Sampling) ship *Resolution* can drill to a depth of 9,000 m. Twelve computer-controlled thrusters keep it in precisely the right spot. Cameras and sensors to take measurements can be dropped onto the seabed. The Bathysnap camera floats back to the surface automatically after taking pictures.

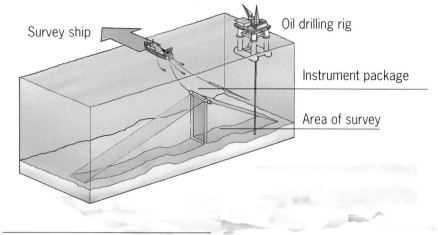

FACTFILE

In the remotest areas of the American 'Wild West', people who drilled new oil wells in an unexplored area often had to kill the wildcats that lived there. The dead animals were hung on the derrick (drilling tower). So, wells that were sunk to find new oilfields became known as wildcat wells and the oil men who drilled them were called wildcatters.

Survey ship

Oil drilling rig

Instrument package

Area of survey

Print out

GLORIA (Geological Long Range Inclined Asdic) produces pictures of the seabed. The instrument package is towed behind a boat. It sends out pulses of sound which are reflected by the seabed. Receivers pick up the reflections and use them to create a picture of the shape of the seabed.

Sandstone and mudstone form multi-coloured strata (layers) of rock in Wales. These rocks were laid down in the Silurian Period, from 439 to 408 million years ago, at a time when the first plants were beginning to grow on the land.

Below: Powerful machines now do much of the heavy cutting, lifting and moving work in coalmines that used to be done by hand. The miners wear electric lamps on their helmets to light their way.

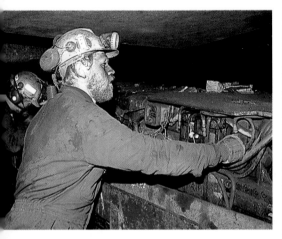

Below: An American miner inspects an auger, a type of drill used to bore into a coal face to extract the coal. As the auger rotates, the sharp 'picks' around its rim bite into the coal.

Coalmining

In early mines, all the work was done by hand, but modern coalmines are full of machines. They cut the coal, hold up the roof, hoist the coal to the surface and transport miners to and from the coal face. Even so, mining is still hard, dangerous work.

Open-cast and deep mining

Today, the way coal is mined depends on how deep it is. If it is within about 60 metres of the surface, it is dug out from a surface mine, or open-cast mine – a huge shallow hole in the ground. The rock containing the coal is blasted with explosives.

When coal lies close to the surface of the land, soil and rock are stripped away so that the coal is exposed and can be dug out. This German coal mine is a surface mine.

The shattered rock and coal are then scooped up by huge dragline excavators or power shovels. Deeper coal deposits require underground mines, or deep mines. A shaft is dug down into the ground to reach the coal. Most coalmine shafts are around 90 metres deep, but the deepest can go down more than 600 metres. Miners dig outwards from the main shaft using cutting machines. After the coal has been cut, it is separated from the rock, crushed and washed.

FACTFILE

The main coalmining methods are longwall and room-and-pillar. In longwall mining, two tunnels are dug through the coal. A machine moves back and forth between the tunnels, cutting the coal. As the cutting machine moves through the coal, the roof behind it is allowed to collapse. In room-and-pillar mining, pillars of coal are left to support the roof.

Gas comes ashore by pipeline from the North Sea gas fields at British Petroleum's North Sea Gas Terminal at Easington in Yorkshire.

Drilling for oil and gas

Oil and gas are extracted by drilling down through the Earth's crust to reach them. A sharp-toothed drill bit attached to a pipe is rotated so that it cuts a hole through the rock. As the drill pipe disappears into the ground, more sections of pipe are screwed on behind it. Aiming a drill used to be haphazard, but engineers can now bend a pipeline this way or that to steer it through rock in any direction. By the time the drill reaches the oil or gas, the pipeline may be several kilometres long.

The turbo-drill

The drill bit used to be rotated by an engine at the surface, so the whole length of the drill pipe had to rotate. Nowadays, it is more common to use a turbo-drill. With this method, the drill pipe itself does not rotate. Chemical mud pumped down the pipe to cool the bit and flush out rock fragments spins a turbine. The turbine drives the drill bit.

Oil is trapped underground at high pressure. When the drill breaks through the rock above it, the oil can rush up to the surface and spurt out of the top of the drill pipe. These 'blow-outs' or 'gushers' used to be very common, but nowadays there are special valves at the well-head, called blow-out preventers, that stop them.

When there is not enough pressure to push oil to the surface, water or gas can be forced into the well (step 1) to push the oil out (step 2). This 'nodding donkey' is injecting steam into an oil well to increase production.

Drill ships are used in the deepest waters, up to 2,400 m. At such depths, anchors cannot be used to hold the ship steady. Instead, the ship positions itself with great precision using navigation satellites and directional propellers called thrusters.

Offshore oil rigs

About one-third of the world's oil comes from oilfields underneath the seabed. Oil is extracted from them by offshore production platforms, or rigs. A modern production platform is so huge that it can stand on the seabed 400 metres below. Floating rigs can work to a depth of 1,000 metres. Platforms as big as skyscrapers anchored to the seabed drill wells 5,000 metres deep. A large platform may weigh 200,000 tonnes and store a million barrels of oil in storage tanks around its legs.

Right: A production platform stands off the coast of Norway. Its crane towers over the accommodation block that houses the platform's workers. On top of the block is a helipad, for ferrying workers ashore by helicopter.

Satellite fields

A really big field may have several production platforms sitting above it, but many oilfields are too small to have their own platforms. In the past they were ignored because of the high cost of developing them. Nowadays, small fields are developed by linking them to nearby production platforms by underwater pipelines. These small outlying oilfields are called satellite fields.

Offshore drilling for oil usually involves anchoring a drilling rig to the seabed. An alternative in shallow waters is to build an artificial island, as was done here in the Beaufort Sea in Canada's North-West Territory.

F A C T F I L E

The main offshore oilfields are beneath the Arabian Gulf, the Gulf of Mexico and the North Sea between Scotland and Norway.

Piping fuel ashore

The North Sea is one of the world's major oil- and gas-producing regions. It is also one of the most difficult parts of the world to drill for oil. The offshore platforms that bring the oil to the surface stand in water up to 180 metres deep, lashed by waves as high as 30 metres and winds blowing at 120 kilometres an hour. Large oil and gas fields were discovered in the North Sea after a major gas field found in 1959 attracted oil-exploration companies to the area. The first oil discovery came in 1969. At first the oil and gas were brought ashore by tankers, but now much of it is piped ashore through more than a thousand kilometres of pipeline.

Below: Oil brought to the surface by fixed and semi-submersible platforms and remote sub-sea manifolds (feeder pipelines) is held in storage tanks until oil-tankers can take it on board from floating filling towers.

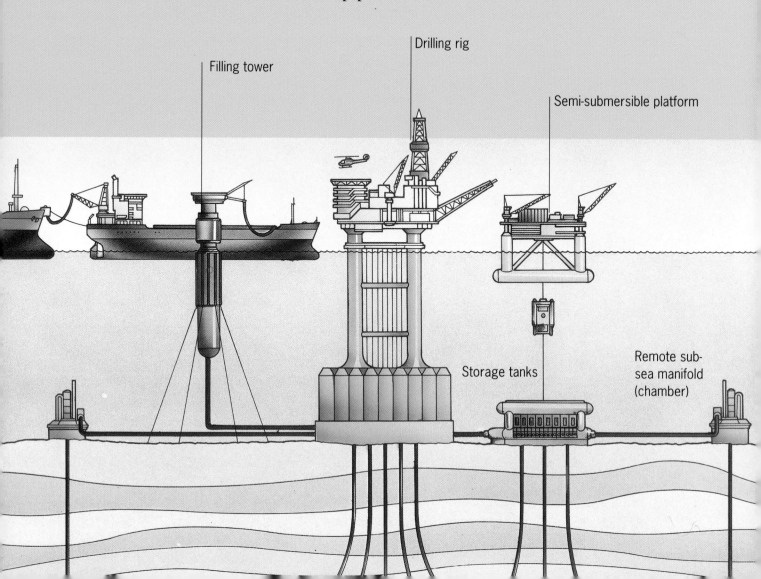

Filling tower

Drilling rig

Semi-submersible platform

Storage tanks

Remote sub-sea manifold (chamber)

Looking for new fields

The oil prospectors are now seeking new fields to the west of Scotland. If the North Sea is difficult, the western seas are even worse. The rigs will have to work in water up to 1,000 metres deep, battered by waves that have travelled all the way from the North American coast, gathering energy all the way. Floating rigs will have to be used instead of rigs that rest on the seabed.

The Ekofisk field was the first major oil discovery in the Norwegian sector of the North Sea. This large oilfield was found at the end of 1969. In the 1970s, more oilfields were found here.

Refining crude oil

The oil that comes out of the ground is called crude oil. It is a mixture of different liquids, gases and solids and it looks like thin black treacle. In this form, the oil is useless, but it can be changed into a wide range of useful products by processing it in an oil refinery. At the refinery, the crude oil is separated into its different parts by using heat, pressure and chemicals. A refinery transforms crude oil into petroleum gases, gasoline (petrol), kerosene (jet fuel), diesel oil for trucks, lubricating oils, waxes and bitumen (tar).

There are about 1,000 oil refineries in use in the world today. About a quarter of them are in the United States. This one is in Puerto Rico.

Below: An oil refinery contains hundreds of kilometres of pipelines carrying oil and its various constituents between reaction vessels and storage tanks.

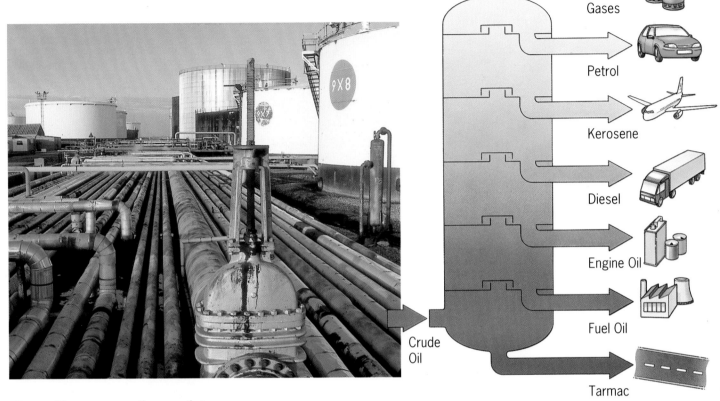

Gases

Petrol

Kerosene

Diesel

Engine Oil

Fuel Oil

Crude Oil

Tarmac

Distillation and cracking

The two main ways in which oil is processed in a refinery are distillation and cracking. The first step is to heat the oil in a fractional distillation tower. The lightest oils and gases rise to the top of the tower, while heavier, thicker oils, waxes and tar stay near the bottom. Different materials leave the tower at different heights. Crude oil separation produces too much of the heavier oils and not enough petrol, so various chemical processes are used, called 'cracking', to increase the amount of petrol produced.

In addition to gases and oils, a refinery also produces a range of chemicals that the chemical industry uses to make plastics, detergents, fibres, medicines, insecticides, weed-killers and many other materials.

A distillation tower uses heat to separate crude oil into different materials. The heat drives lighter materials higher up the tower, leaving heavier oils, waxes and tar at the bottom. The types of fuels and everyday products formed from the oils, waxes and tars is shown.

The oil refinery

At first sight, an oil refinery is a bewildering muddle of pipes and tanks, but it is actually laid out in a carefully planned and logical way. Dozens of oil tanks – known as a 'tank farm' – receive oil from tankers docking alongside them. From there, the oil is piped to distillation towers, first at atmospheric pressure and then at high vacuum (complete absence of air). Then a series of cracking units break down the oil into simpler substances by heat (thermal cracking) and by chemicals (catalytic cracking).

The various products from these cracking units are piped to other parts of the refinery where they are treated further. Blowing air from the distillation towers through the bitumen forms different grades of bitumen, from soft to hard and brittle. Heavy oils are chilled in refrigerated tanks so that the waxes contained in the oils can crystallize and then be removed. Finally, the products are put in storage tanks before being transported to their users

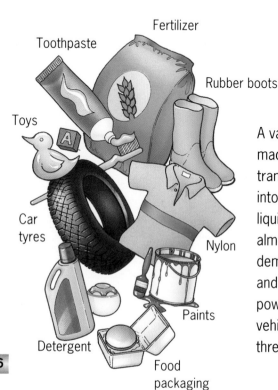

A variety of household items are made from crude oil. Oil refineries transform thick black crude oil into an amazing variety of gases, liquids and solids that are used in almost every industry. The most demand is for petrol, kerosene and diesel oil, the fuels that power the world's engines and vehicles. They account for about three-quarters of all oil products.

FACTFILE

The world's largest oil refinery is the Amoco refinery in Texas City, Texas, USA. It can handle more than 400,000 barrels of oil. A standard oil barrel is 159 litres, so the refinery's capacity is equivalent to 64 million litres of crude oil.

The sun lights up the morning sky at the Fredericia oil refinery in Denmark. Refineries are lit up like Christmas trees at night because they work 24 hours a day to meet the demand for their products.

Fleets of tanker lorries carry the refinery's products away by road to chemical plants, fuel stations and other users. Products are also collected and sent to chemical plants by rail, using tanker wagons.

Moving oil around the world

Oil is transported around the world through surface and underground pipelines and in giant tanker ships. The business of moving oil is enormous. At any time, almost half of the cargo on the world's oceans is oil. And the ships are huge. The biggest tankers carry 400,000 tonnes of oil. The ship is divided into a series of separate tanks, to stop the oil for sloshing about too much and rolling the ship over.

Pipelines

On land, the easiest way to move oil is to pump it through a pipeline. Because the oil is thick, the pipes are wide, perhaps a metre across. In the coldest parts of the world, the pipeline has to be insulated to keep the oil warm enough to keep it moving. Gas fields near land are linked to onshore gas terminals by pipelines, but gas from more remote fields has to be transported by tankers. It is purified and then liquified by cooling it to below minus 161.5 °C, so that it occupies a smaller volume, before being pumped into specially designed tankers.

A supertanker offloads some of its cargo of oil into a smaller tanker that can enter an oil terminal and unload the oil into storage tanks.

Below: Crude-oil tankers are simple cargo ships, as shown. Natural gas that cannot be piped ashore is transported by LNG (liquid natural gas) tankers. The gas is changed into a liquid and stored in spherical containers inside the ship.

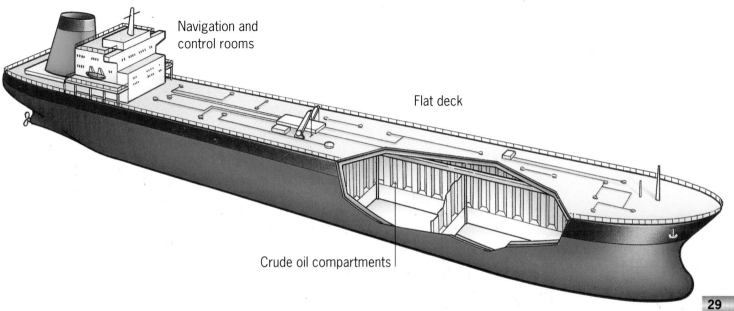

Navigation and control rooms

Flat deck

Crude oil compartments

Left: Serious air pollution has been caused in Europe by the widespread burning of lignite in the former German Democratic Republic (east Germany or GDR). Lignite is a poor fuel, but the GDR had large reserves and so it was used to reduce expensive fuel imports.

A blue smog haze hangs over Hong Kong. A weather condition called a thermal inversion can make air pollution worse by holding it over a city. It happens when a layer of cool air containing the pollution becomes trapped underneath a blanket of warm air.

The effects of fossil fuels on the environment

When coal and oil are burned, they give off gases that are harmful to health and damaging to the environment. Pollutants from fossil fuels burned in power stations and road vehicles include sulphur dioxide, nitrogen oxides, carbon dioxide and carbon monoxide. They irritate the lungs, make breathing difficult, cause acid rain and contribute to global warming (warming of the Earth's atmosphere).

When sulphur dioxide or nitrogen oxides mix with moist air, a chemical reaction between them produces acid. Then, when it rains, it rains acid! But before the rain falls or the moisture settles, the gas or the clouds may be blown hundreds of kilometres by the wind. Acid rain has damaged forests, buildings, lakes and rivers across Europe, Scandinavia and North America.

Smog is a choking cloud of smoke and fog that was common when people heated their houses by burning coal. Smog still happens today, but for a different reason. When chemicals produced by petrol engines mix with air in strong sunlight, the result is a murky haze called photochemical smog. If weather conditions trap this smog over a city, people suffer breathing problems and some may even die.

F A C T F I L E

"Photochemical smog" over Los Angeles in the USA can cut visibility to less than a kilometre. As a result, the State of California has imposed the world's strictest restrictions on car exhaust levels.

Oil floats on water, so a small amount of spilled oil spreads out to cover a huge area of the sea. On 15 February 1996, the *Sea Empress*, an oil-tanker carrying 131,000 tonnes of North Sea crude oil, ran aground as it approached the Milford Haven oil terminal in Wales. About 6,000 tonnes of oil spilled into the sea and washed ashore. Tugboats tried to hold the tanker in position using cables, but the cables snapped in gales and the ship ran aground again. By the time tugs could bring the ship under control, six days later, and start pumping out its tanks, 73,450 tonnes of oil had been lost into the sea.

Cleaning up the mess

Spilled oil can be dispersed by detergents. It can be scraped off beaches or blasted off rocks by water jets. Floating booms around a tanker can stop the oil from spreading in calm weather. Machines can skim oil off the water's surface and collect it in tanks. Oil still inside the ship can be offloaded into another tanker. Eventually, natural micro-organisms break down the remaining oil.

FACTFILE

The worst spill from an oil-tanker released 280,000 tonnes of oil into the Caribbean Sea after a collision between two giant tankers, the *Atlantic Empress* and the *Aegean Captain*, in 1979.

Workers ankle-deep in crude oil tackle the unpleasant task of removing oil, spilled by the *Sea Empress*, from Welsh beaches. This is far more difficult and time-consuming work than dealing with oil-slicks at sea.

Effects on wildlife

Sea birds and seals contaminated by oil can be cleaned, but few heavily oiled animals survive, especially if they have eaten the oil. Shellfish and all the other tiny creatures that live on the seabed can be wiped out by the poisonous effects of the oil or the detergents and chemicals used to remove it.

Tugs try to hold the oil-tanker *Sea Empress* steady after it grounded on rocks that ripped holes in its hull. They eventually lost their battle in the teeth of gales that swept across the area.

Engines

Fossil fuels are widely used in the world today because, when they burn in air, they release the energy stored inside them very easily. Burning, or combustion, converts the stored chemical energy into heat energy. The heat produced makes gases expand rapidly and powerfully. It is this expansion of gases that is harnessed by the engine to do useful work.

The car engine

The latest car engines, like this Nissan prototype, have their own on-board computers. These ensure that fuel is burned in the most efficient way, to produce the most power and the least pollution.

Inside a car engine, petrol and air are drawn into a cylinder, squeezed and then ignited. The burning gases expand, pushing a piston down a cylinder. This downward thrust is converted into the spinning motion of a shaft that drives the car's wheels. The hot gases escape through the exhaust.

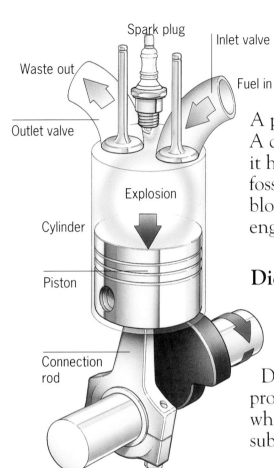

Spark plug

Inlet valve

Waste out

Outlet valve

Fuel in

Explosion

Cylinder

Piston

Connection rod

Crankshaft

A petrol engine needs a spark to ignite the fuel. A diesel engine compresses the fuel so tightly that it heats up enough to burn. A jet engine also burns fossil fuels, but instead of the suck, squeeze, bang, blow series of individual explosions inside a car engine, a jet engine burns fuel continuously.

Diesel-electric drive

Sometimes, a fossil fuel engine does not drive a vehicle directly. Many railway locomotives and submarines are diesel-electric vehicles.
Diesel engines drive electricity generators to produce electricity. This powers electric motors which turn the locomotive wheels or the submarine propeller.

Above: In a car engine, an electric spark ignites the fuel. This pushes the piston down, which in turn rotates the crankshaft linked to the driving wheels.

Below: A Boeing E-3 Sentry aircraft takes off, leaving black trails in the wake of its four jet engines. Newer jet engines are cleaner and so better for the environment.

The air inside a hot-air balloon is heated, by burning propane gas, to make it expand and become lighter than the surrounding air. Propane is one of the gases produced in an oil refinery.

Everyday needs

We use fossil fuels every day in our homes and at work. Natural gas is used for cooking and heating. Gas-fuelled incinerators destroy some of our rubbish and industrial waste by burning it at a very high temperature. Even the electricity we use depends on fossil fuels, because much of it is made in power stations that burn coal or oil.

Portable power

Gas is a particularly useful source of energy because a lot of gas can be compressed into a tiny volume inside a tank. So, gas is more portable than coal or oil. Campers, mountaineers and explorers can take bottled gas anywhere with them. Hot-air balloons carry bottles of gas to heat the air inside the balloon and make it rise. Gases such as propane and butane, which are used for balloons and to pressurize aerosol sprays, are made from crude oil.

FACTFILE

A gas burner called a Bunsen burner is used in science laboratories. The temperature of its flame can be varied by turning a collar on the gas pipe to let more or less air in to mix with the gas.

Above: Barbecues are a popular means of cooking food at home outdoors in many parts of the world. Portable barbecues usually burn charcoal, but fixed barbecues often have a piped supply of a gas such as butane.

White clouds of condensing water vapour rise from the cooling towers at Ferrybridge oil-fired power station in Yorkshire, UK. The special shape of the towers sucks air upwards, cooling the warm water that circulates through them.

Power stations

Power stations are the biggest users of fossil fuels. In a coal-fired power station the coal is pulverized, or turned into dust. By increasing the surface area of the coal in this way, it burns much more quickly and completely. Hot air blows the coal dust into a furnace where it burns more like a gas than a solid. The heat generated turns water to steam which drives electricity generators.

How much fuel?

A modern power station furnace can burn 300 tonnes of coal dust per hour. An oil-fired power station can burn 115 tonnes of oil per hour – 2,750 tonnes per day. So, coal-fired power stations are usually built near coalfields and oil-fired power stations are usually built near oil refineries. Some power stations are built to run on all the main fossil fuels as well as on peat, for example, in Ireland and on orimulsion, another crude oil product.

Power station efficiency

Fossil fuel power stations are very inefficient. The best of them waste two-thirds of the energy from the fuel – most of it escapes as heat. Combined heat and power stations are more efficient because they use the heat as well as the fuel. Many Scandinavian and European cities – Copenhagen, Stockholm, Berlin, Munich, Milan, Oslo and Paris among them – have heat mains laid underground to distribute heat from power stations to homes and businesses.

F A C T F I L E

About three-quarters of all the fuels used throughout the world are fossil fuels. Oil is the most widely used fuel, accounting for nearly 31 per cent of the world's fuel consumption, followed by coal (26 per cent) and natural gas (19 per cent).

A fossil fuel power station is like a giant steam engine. The fuel is burned to make steam, which spins turbines, and the turbines drive electricity generators.

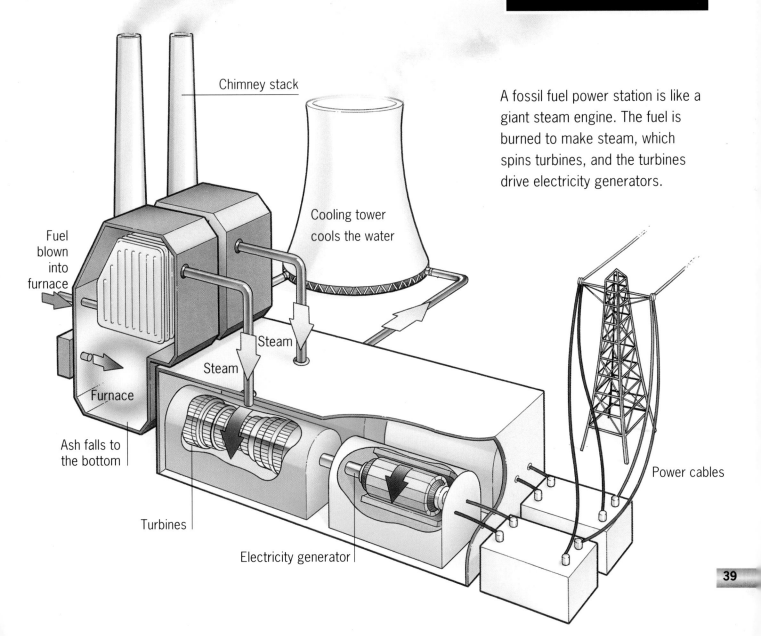

Chimney stack

Cooling tower cools the water

Fuel blown into furnace

Steam

Steam

Furnace

Ash falls to the bottom

Turbines

Electricity generator

Power cables

The Danish gas network

In the late 1970s, politicians wanted to build nuclear power stations in Denmark to reduce the country's need for expensive oil imported from the Middle East. People protested in such large numbers that the government abandoned the plan. But the energy had to come from somewhere.

In 1979, work started on a network of pipes to bring natural gas from the North Sea to people's homes. Burning gas, which is a cleaner fuel than either coal or oil, has reduced air pollution. In the first ten years, sulphur dioxide emissions were 235,000 tonnes lower than if the same energy had been supplied by burning coal and oil. Nitrogen oxide emissions have been reduced by 26,000 tonnes and carbon dioxide emissions by 13.7 million tonnes.

Below: Denmark is being criss-crossed with a network of pipelines to carry natural gas from the North Sea to all parts of the country.

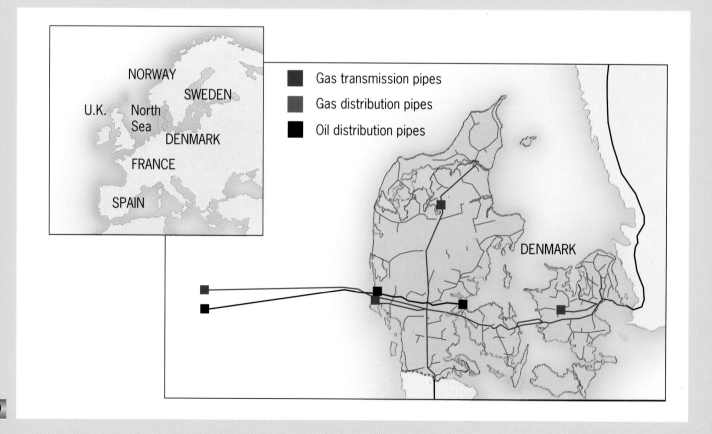

NORWAY

SWEDEN

U.K.   North Sea

DENMARK

FRANCE

SPAIN

■ Gas transmission pipes

■ Gas distribution pipes

■ Oil distribution pipes

DENMARK

Coal power

Denmark's second source of energy is coal. But although coal-fired power stations have helped fill Denmark's energy gap, they have also created a lot of pollution. Burning coal produces more carbon dioxide, a greenhouse gas, than any other fossil fuel.

Greenhouse gases are responsible for making the Earth warmer and even changing its climate. So Denmark is now trying to reduce its dependence on coal by developing wind power.

Construction work continues on Denmark's natural gas distribution network. Water crossings were a particular challenge in the strong currents of the inner Danish seas.

The Studstrup power station is one of Denmark's biggest electricity- generating plants. It burns coal most of the time, but it can burn oil too. It also pipes hot water to homes and industry within 30 km of the power station. About 8 per cent of the Danish population, 400,000 people, receive heat from the Studstrup power station.

In future, coal could be delivered as a liquid. When coal is mixed with water and special chemical additives, it can be transported by pipelines or tankers just like oil. And when it is sprayed as a fine mist of droplets, burning it leaves only small amounts of ash.

Cutting pollution

Fossil fuels are a valuable source of energy, but if they are to be used in future, ways will have to be found to reduce the pollution they cause. Burning coal causes more pollution than burning oil or gas, so more work goes into preventing pollution from coal. Power station chimneys can be fitted with equipment that removes a lot of the more harmful pollutants, such as sulphur.

Combined heat and power systems also – indirectly – cause less air pollution. They use their hot waste gases to heat water, which is then used to heat buildings. This both saves on fuel and, by burning less of it, reduces the amount of carbon dioxide released into the air.

Tackling traffic pollution

One way of reducing pollution from vehicles is to design new engines that are more efficient so that they burn fuel more completely and allow less unburned fuel to escape. All petrol used to have lead added to it to make it burn more smoothly and evenly. More and more cars are now fitted with engines that burn unleaded petrol.

The change to unleaded petrol will continue. More cars will have catalytic converters, which remove pollutants such as nitrogen and sulphur oxide – though not carbon dioxide – from their engine exhausts. And laws will force more manufacturers to make zero-emission cars, such as solar-powered electric cars, which cause no pollution. Laws will also force drivers of petrol-driven vehicles not to use them in towns and cities. Public transport will be promoted and government funded.

Above: A catalytic converter, a device on a car that changes up to 90 per cent of the poisonous gases produced by burning petrol in the engine into relatively harmless ones.

Below: Filling a car with unleaded petrol. The unleaded petrol available today is cleaner and safer than the old leaded petrol Lead is a poisonous pollutant when released into the air.

A car is filled with liquid hydrogen fuel. The hydrogen is obtained from water by splitting the water molecules using electricity. Photovoltaic cells – special glass units that trap the energy of sunlight and convert it into electricity – provide the necessary power.

New fuels

Hydrogen is a much cleaner fuel than coal or oil. When it burns, it combines with oxygen in the air to form harmless water vapour. Hydrogen can be made from fossil fuels. Perhaps, in future, we might use fossil fuels by first converting them to a cleaner fuel, such as hydrogen, inside special conversion plants where harmful pollution can be dealt with before it can escape into the air. Several car manufacturers have already built experimental hydrogen-powered cars.

Hydrogen could be made in vast quantities by splitting water into hydrogen and oxygen using solar and wind power. But converting fossil fuels might be the first step towards the more widespread use of hydrogen. One day, fuel stations might have hydrogen pumps alongside their petrol pumps.

A technician assembles a set of fuel cells. A chemical compound that releases oxygen is added to the cells. The oxygen combines with hydrogen from the fuel, producing electricity.

Below: Diagram of a fuel cell. The only waste product is water. Oxygen and hydrogen pumped into the cell combine to form water, and electricity is produced.

H_2

H_2O

O_2

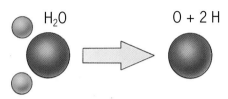

H_2O → $O + 2 H$

A molecule of water splits into one atom of oxygen (red) and two atoms of hydrogen (blue).

Fuel cells

The few electric cars on the streets today use heavy batteries as a source of power. Future electric cars could be powered by fuel cells. Unlike batteries, which store chemical energy, fuel cells actually make electricity when it is needed, by combining hydrogen and oxygen. The only by-product is water, so fuel cells are a very clean way of making electricity. Hydrogen made from fossil fuels could be used to make electricity in fuel cells.

GLOSSARY

Anthracite The oldest and hardest type of coal.

Barrel A standard measure of oil equivalent to 159 litres.

Billion One thousand million.

Bitumen A thick black tar-like material found naturally or made from crude oil.

Catalytic converter A device fitted to a petrol engine that chemically converts harmful gases produced by the engine into less harmful gases.

Cracking Processes using heat and chemical reactions that help to break down crude oil and convert it into more useful materials.

Crude oil Natural oil as it comes out of the ground, before it is refined.

Derrick The tower holding the drilling machinery that stands over an oil well.

Distillation The process of purifying a substance by heating it so that it evaporates, and then cooling it so that it changes back into a liquid.

Energy The ability to do work.

Ethane A flammable gas found in natural gas.

Flare The flame produced when an oil rig burns off unwanted gases that rise from an oil well to the surface along with crude oil.

Fossil fuel A fuel such as coal, oil or natural gas formed from the remains of microscopic plants and animals that lived millions of years ago.

Fuel A material that is burned to produce heat or generate electricity.

Fuel cell A device that makes electricity by means of a chemical reaction between hydrogen and oxygen.

Generator A machine designed to change the movement energy of a spinning shaft into electricity.

Global warming Warming of the Earth's atmosphere by gases such as carbon dioxide trapping the Sun's heat.

Greenhouse gas Any gas, such as carbon dioxide, that traps the Sun's heat and so contributes to global warming.

Hydrocarbon A material that contains the chemical elements hydrogen and carbon linked together chemically.

Hydrogen The lightest gas and a useful fuel, found in water and in crude oil. It burns readily in air.

Kerosene A fuel oil used in jet engines, made from crude oil.

Lignite The softest and youngest coal, formed perhaps only one million years ago.

Megawatt (MW) One million watts, a measure of electrical power.

Methane A flammable gas found in natural gas, also called marsh gas.

Natural gas Gas found in nature, usually in deep underground pockets, often with crude oil.

Oil rig An offshore platform used for drilling for oil or for pumping it up.

Petroleum Another name for crude oil.

Power station A building where energy from a fuel is used to make electricity.

Sea coal Pieces of coal washed out of underwater coal deposits and carried ashore by the sea.

Sediment Fine particles carried on the wind or by water that collect together and, in time, become compressed to form rock.

Smog A mixture of smoke and fog that collects over some large cities.

Surveying Studying the land, its shape and its rock types in the search for valuable fuels and minerals.

Tar sands A soft rock, such as sandstone, that has soaked up oil. The oil can be extracted from the rock.

Turbine Angled blades fitted to a shaft that is free to rotate. A moving gas or liquid pressing against the blades makes the turbine rotate.

Books to read

Cycles in Science: Energy by Peter D. Riley (Heinemann Library, 1997)
Earthcare: Raw Materials by Miles Litvinoff (Heinemann Library, 1996)
Eyewitness Science: Energy by Jack Challoner (Dorling Kindersley and London Science Museum, 1993)
Science Topics: Energy by Ann Fullick and Chris Oxlade (Heinemann Library, 1998)
The Greenhouse Effect by Alex Edmonds (Franklin Watts, 1996)
Science Works: Energy by Steve Parker (Macdonald Young Books, 1995)
The Super Science Book of Energy by Jerry Wellington (Wayland, 1994)
The World's Energy Resources by Robin Kerrod (Wayland, 1994)

Power and energy consumption

Power is the measurement of how quickly energy is used. It is measured in joules per second, or watts. An electric iron might need 1,000 watts to work, but a portable radio might need only 10 watts. The energy needed to keep the radio going for one hour would run the iron for only six minutes, because the iron uses up energy ten times faster than the radio. The diagram to the right compares the power ratings of household electrical goods and of homes and power stations.

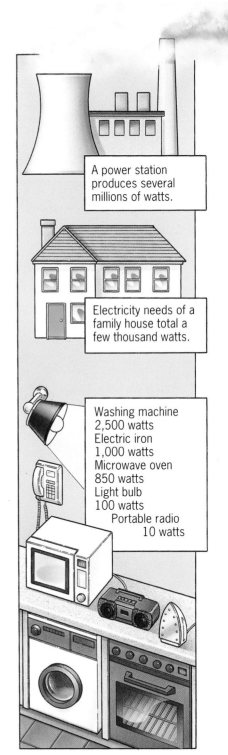

A power station produces several millions of watts.

Electricity needs of a family house total a few thousand watts.

Washing machine 2,500 watts
Electric iron 1,000 watts
Microwave oven 850 watts
Light bulb 100 watts
Portable radio 10 watts

INDEX